All Scripture references from the KJV of the Holy Bible, unless otherwise indicated.

Prayers Against Barrenness for Success in Business and Life, by Dr. Marlene Miles

Freshwater Press, 2023

ISBN: 978-1-960150-82-0

Paperback Version

Copyright 2023 by Dr. Marlene Miles

All rights reserved. No part of this book may be reproduced, distributed, or transmitted by any means or in any means including photocopying, recording or other electronic or mechanical methods without prior written permission of the publisher except in the case of brief publications or critical reviews.

Thank You for purchasing and reading this book. May it bless you richly, grow you mightily and may you feel the Lord's presence always.

Shalom,

Dr. Marlene Miles

Table of Contents

Barrenness 5
The Curse of Barrenness 15
Free Your Mind: Deliverance *of the* Head 17
Cursed Hands 24
Dry Hands 33
Good Help Wanted 40
Lazy Hands Will Be Barren 47
Marital Barrenness 52
It Could Be My Fault 53
Wealth Transference 55
Cursed Legs 60
Fire Back 66
Ancestral Wickedness 68
Loose All Bondage 73
Blessings of Abraham 82
Household Witchcraft 84
Deliverance & Destiny Helpers 86
Angelic Assistance 88
The Thieves 91
Power In the Hands 95
Christian books by this author 97

Prayers Against
BARRENNESS
for Success in Business & in Life

Freshwater Press, USA

Barrenness

Everything you do, everything you set your hands to do shall prosper. That's what the Word says and for that to become your reality, you need blessed hands, among other things.

The *spirit of barrenness* can affect more than having children, it can steal a man's dreams, and his whole life. There is such a thing as scholastic barrenness. As a student who may have just failed an exam, you may be saying to yourself, I know this stuff, what went wrong? You're super smart--, book smart. What happens on exam day? Why didn't you score well on the test? Freezing on exams or when the teacher asks you a question, or when having to speak publicly, as if you know nothing, when you know this material perfectly, could be signs of barrenness working against you to bring you to shame, failure, and disappointment. Setting your mind and hands to do things, but they constantly don't work out means you need to find out if you're under judgment from God, or what exactly is going on, since evil human agents love to see people fail exams.

Business barrenness can be caused by iron earth and or brassy Heaven--, both are under the Curse of the Law. There are other causes for barrenness in career and business and much of this book is about and praying about just that. We want the successful life God says we should have.

Lord, I worship You; You are great and greatly to be praised. Please forgive me for anything that would stand against my prayers, in Jesus' Name.

Lord, help my obedience so You will be well-pleased with me.

I repent for and renounce any idols in my life, whether I created them, or simply accepted them, in the Name of Jesus.

I refuse to bow down to idols, or any of the things You frown upon that bring barrenness into my life, in Jesus' Name.

Father, enable me to observe the Sabbath and have reverence for the sanctuary of God; You are the Lord.

Lord, help me follow Your decrees and obey Your Commands, in Jesus' Name.

Lord, send Your servant rain in season, that my crops and trees will yield abundantly.

I break the curse of living hand to mouth, in the Name of Jesus. Lord, Your Word promises harvesting until the next planting, all the food we want to eat and eating from our store, not our basket, in Jesus' Name.

Lord, thank You for harvests so abundant that there won't be room enough to receive it all, to store it all, our barns bursting forth with plenty, that we have enough to bless others, in the Name of Jesus.

Lord, I break the curse of business *barrenness* by Fire and by Force, in Jesus' Name.

Lord, let all my enemies fall by the sword; let me see the destruction of the wicked that rise up against me, in the Name of Jesus.

The favor of the Lord is life; Lord thank You for divine favor, in the Name of Jesus.

Lord, make me fruitful, productive, and abundant, in school, in business, in life, in the Name of Jesus.

Covenant-keeping God, let me be like You, keeping laws, statutes, rules, and Covenant, in the Name of Jesus.

Lord, thank You for increasing our numbers of my business, in my favor, multiplying successes, and making my business strong, in the Name of Jesus.

Lord, tabernacle with me, dwell in me, and I in You, let me be one of Your people, in the Name of Jesus. You are the Lord.

Lord, let me walk with You, and You with me. You are the Lord.

I refuse to plant with no harvest. Enemies of God, you will not destroy my crops, my plans, or my harvests, in the Name of Jesus.

Lord, give me divine sight and vision in education, business, and life, in Jesus' Name.

The joy of the Lord is my strength; every power eyeing my strength, prosperity, or abundance, fall down, and die, in the Name of Jesus.

Lord, do not let my enemies rule over me, in the Name of Jesus.

Spirit of fear, I reject you, I rebuke you, in the Name of Jesus; I will not back down, I will not fear – even though I walk in the valley of the shadow of death, I will not fear for the Lord is with me. He is the Lord.

Spirit of pride, leave me, in Jesus' Name.

False humility, leave me, in Jesus' Name.

Iron sky, I reject you; Lord God, open Heaven over me, in the Name of Jesus.

Bronze ground, I reject you; Lord, make my ground fertile and rich, to bring forth abundantly all that is planted, in Jesus' Name.

Spirit of vain efforts, I reject you, in the Name of Jesus; my crops shall increase 100-fold like Jacob's did, even in the time of famine, in Jesus' Name.

Time, be on my side in the Name of Jesus. Time, work for me, in the Name of Jesus.

Lord, bless the works of our hands, bless my marriage, family, children, businesses, and career, in the Name of Jesus.

Lord, when I need to be corrected, let me receive correction with Wisdom, Grace, and joy, in the Name of Jesus.

Lord, have Mercy that I do not fall under judgment from an angry God. Let me be named among Your *people*, let my worship ascend to You as a sweet-smelling savor, in Jesus' Name.

Give us this day, our daily bread; do not cut off my staff of bread, in the Name of Jesus.

Spirit of dissatisfaction, I bind you, and I *loose* the *spirit of satisfaction,* every day, and for every good gift from the Father, in Jesus' Name.

I repent for and renounce any *idols* in my life, whether I created them, or accepted them, in the Name of Jesus.

Lord, I pray for the prosperity of the city I live in; prosper the land, so that I will prosper, in the Name of Jesus.

Covenant-keeping God, let me be just like You, keeping sabbaths, Holy Days, and appointed times, in the Name of Jesus. You are the Lord.

If not observing the Sabbath is the reason the land is not yielding to me, Lord, forgive me, have Mercy. I will observe Sabbath from here on out, in Jesus' Name.

Lord, let there be worship toward You, and rest for me and the land in which You've placed me, on the Sabbath, in Jesus' Name.

Spirit of fear, I reject you, I rebuke you, in the Name of Jesus; I will not back down, I will not fear – even though I walk in the valley of the shadow of death, I will not fear for the Lord is with me. He is the Lord.

Lord, do not let me fear the ridiculous, in the Name of Jesus – not a leaf in the wind. Do not let me fall, whether I am being pursued by enemies, or not, and especially if I'm not, in the Name of Jesus. Amen.

Give me strength and wherewithal to stand in the evil day, to stand against every enemy of my soul, every enemy who comes up against me, and in the whole armor of God, to stand therefore, in Jesus' Name.

Spirit of wasting away, pining away, spirits of depression and devastation, get far away from me, in the Name of Jesus.

Lord, I confess my sins and the sins of my ancestors—their unfaithfulness and their hostility toward God. He is the Lord.

Lord, remember Your covenant with Jacob and Your covenant with Isaac and Your covenant with Abraham, and remember the land. You are able to save to the utmost; You are the Lord.

Lord, do not let me reject Your laws, statutes and decrees, in the Name of Jesus. Do not let me break Covenant; You are the Lord. You are my God, in Jesus' Name.

Bless the Lord, You've brought me out of sin, out of bondage, out of slavery, You've broken evil yokes and taken evil loads, for Your yoke is easy and Your burden is light; You are the Lord.

The Curse of Barrenness

Being prosperous is a gift of God. Not being prosperous is considered a curse. God has been known to cut off a man's seed. Many times, that meant the man did not have sons, but it could have also meant that he was sterile and could not impregnate his wife at all, so there would be no children of the marriage. Being fertile, in all things, the fruit of the womb as well as in daily life, and in business, Bible folks thought that **proved** that God was with you. Anything else brought reproach. That was their form of keeping up with the Jones's. Shall we say, the *Jonesites* of their time.

Reproach is bad enough, but the frustration of nothing working in your hands is *too much*. Desperation, the depravity of man, devil ideas have led many into devil-inspired shortcuts to get things and stuff, to avoid the leering eye of neighbors and the stigma that comes with it. Problem is, the devil offers shortcuts to bypass God, as the devil *plays* man while he's playing God. This is when sin and curses enter in. If there is unrepented sin, there are curses and iniquity.

Personal sin can cause barrenness. God closed all the wombs of the house of Abimelech for taking Sarah into his house, (Genesis 20:18). Abimelech didn't even sleep with Sarah and there was a curse of barrenness on his house! Don't tempt God.

Michal, David's wife who made fun of his worship of God had no children. God is not mocked, and we do not get to judge anyone's worship. Barrenness is many times a curse from God, so we must seek God to be sure we are not under judgment from God as to why we are not prospering in anything from business to the bedroom.

Free Your Mind: Deliverance *of the* Head

A woman in the Bible lost a coin and searched everywhere in the house for it. Where did you put that thing of value in your life? Where did you put that key? Where did you put that money? You have to be clear-minded, clear-headed, and in charge of your faculties when you are conducting your life, working, cleaning, organizing. When you are placing precious things in a certain place, you must remember.

Memory is a gift of God. Do not let the enemy capture your memory. Do not lose focus; **any power after your head is after your entire body and your life**.

Do not be distracted when you are doing something significant. Our God is a God of order. Successful people most often have a place for everything and everything is in its place. If you use something, put it back in its place. If you move something be sure you remember it's new location. It is best not to be distracted when organizing things, especially moving things to new locations, and/or changing passwords.

If you are not one to write down a new password because you have a great memory, get off the phone with your BFF dishing the dirt while simultaneously changing your password to include at least 8 characters, a lower case, an upper case and a special symbol that you haven't used the last 10 times. Even if you have a great memory, it's as though the conversation will steal away that **PErfectpassword12*** you just made up. Hang up the phone, focus, remember, *or* write it down.

As with the woman with the coin, *found* money in your house would not have ever been lost if you had focused on what you were doing when you put it in that special place for safe keeping.

You're welcome.

If your memory is under attack, you need deliverance of the head. The demonic things that bring *barrenness* often begin with attacks on the mind, the head, in attempts to bring the head into captivity. Constant headaches or weird feelings in your head need to be dealt with immediately. If the doctors can't find anything wrong with you, most likely these are evil arrows.

I cover myself with the Blood of Jesus.

My head, reject every evil arrow. Back to sender, in the Name of Jesus.

Evil hands assigned to capture my head, or my mind, wither, in the Name of Jesus.

Every arrow of untimely death, fired into my brain, miss your target, return to sender, in Jesus' Name.

Any evil covering cast or veil on my head, catch fire, in the Name of Jesus.

Every curse operating against my head, die by the power in the Blood of Jesus.

Every manipulation of my glory through my hair, scatter now, in the Name of Jesus.

Every power using my hair against me, backfire, and then die, in Jesus' Name.

Every hand of the strongman upon my head, wither and become impotent, in the Name of Jesus.

Every power of death, hell, or the grave, up against my head, die, in Jesus' Name.

Holy Ghost Fire, consume every satanic deposit in my head, in the Name of Jesus.

My head, receive deliverance by Fire, in the Name of Jesus.

O God, arise, thunder in Your power and Your majesty and scatter my tormentors, in the Name of Jesus.

Power of God, arise, destroy to desolation all covens assigned against my head, in the Name of Jesus.

O God, arise, raise up a standard and draw me out of the proud waters, in Jesus' Name.

Every incantation of evil by the power of darkness in the spirit realm against my head, I nullify you, by the Blood of Jesus.

Rain of Wisdom, knowledge and favour, fall upon my head, in the Name of Jesus.

Voices of strangers casting spells against my head, be silenced, die, in Jesus' Name.

Blood of Jesus, Living Water, Fire of God, clear and cleanse my head, in Jesus' Name.

Arrows of darkness fired into my head, back to sender, in the Name of Jesus.

Power of household wickedness upon my head, die, in the Name of Jesus.

My head, wake up by Fire, and **stay woke**, in the Name of Jesus.

Arrows of memory loss, backfire, in the Name of Jesus.

My head, reject and eject all bewitchment and cobweb arrows, in the Name of Jesus.

I negate and nullify the power of every strand of every cobweb set, sent or launched against me, in the Name of Jesus. I cancel the power of all curses upon my head-, back to sender, in the Name of Jesus.

Every evil witchcraft name given to me be dissolved from my forehead and naval, in Jesus' Name.

Any spiritual bat, and/or spiritual lizard introduced into my head receive the Fire of God, come out and die, in Jesus' Name.

Any power calling my head for evil, Blood of Jesus, answer for me, in Jesus' Name.

Every witchcraft arrow fired at my head, back to sender, in the Name of Jesus.

Every evil hand laid upon my head since I was born, die, lose and reverse all power against me by the Blood of Jesus.

Every word curse spoken by parents or relatives working against the glory of my head, break, in Jesus' Name.

Anything stolen from my head when I was a child, I take it back, in the Name of Jesus.

Every power assigned to lock my head in a cage, receive angelic slap, in Jesus' Name.

Invisible loads of darkness upon my head, catch fire, in the Name of Jesus.

Any problem brought to my life through head attacks, die, in the Name of Jesus.

Lord, You've brought me out of sin, bondage, slavery, broken evil yokes and taken evil loads. Your yoke is easy, and Your burden is light; You are the Lord.

Lord, when I was bowed down by sin, when my head was bowed down; You are the lifter of my head; You are the Lord.

Cursed Hands

If you've got a little *destructo* kid – one that can tear up a room in a hurry, there's a reason he or she breaks everything. It could be cursed hands. With your kids or adults, do you hear things being dropped and falling to the floor or ground all the time? I do. Most often at work. Cursed hands.

In work environments machines are expensive, appliances in the home are expensive, cursed hands drop things all the time. Pray to the Lord that the workers you hire do not have **cursed hands**. If they do pray these same prayers that you pray for your own hands, for the hands of those in your family or who work with, or for you.

Pray against hands that break everything; they are cursed hands. Even hands that don't drop everything, but what they touch breaks. Cursed hands may not break anything in the natural, but in the spirit, all success and profits are lost if they put their hands to any part of a project.

I recently asked a guy about his hands, and he was *pleased* to announce that he breaks everything. He over torques a lug with the lug wrench and breaks one or both of those things. He breaks cell phones, tv remotes, cars. He's kind of proud of it. I told him to pray against cursed hands; he ignored me.

Rough hands, not rough skin, but hands that are not gentle could be cursed hands. Hands that hurt others. I had a masseuse whose hands I can hardly describe. Such a lovely, friendly looking person but once my face was down in that circle of that massage table, in that dimly lit room, and her hands first touched me, I wondered WHAT just placed their hands on my back. These hands were ice cold and felt bumpy like crocodile scales. No, she was not wearing gloves, these were her hands. She wasn't rough in her technique, but she had rough hands. I didn't go back there.

Dangerous hands are hands that hurt others. I went to a hair salon and my regular person was out sick, so I was assigned a stylist. Watching her work, the results from her styling chair were absolutely beautiful, so I agreed to be styled by her that day.

Whether tender-headed or not, do not let that woman touch you. She has no sense of touch. If she just moves a little of your hair with her hands, not even using a comb--, her hands, to see if your color is growing out, it hurts. Every touch by this woman hurts. It is not intentional. I told her about my experience with her, and she said, *Yeah, everybody tells me that*, in a tone that let me know she had no plans to do anything about it.

Having *hands* in sports means you can catch, usually. Having *hands* in your profession usually means you can do a nice job. Having *hands* as a surgeon, for example, means the surgical results will be successful. Having *hands* in dentistry means that, and also that you do not hurt the patient.

If you do not have gentle and effective hands, don't choose a profession where you have to touch **people**. If you must, and you feel you have been attacked demonically or cursed, then pray diligently to God for deliverance before you get licensed.

Hands that hurt the land, the Earth, will not prosper you. If the land (area) where you live is not yielding to you, as in the earth beneath you is as iron, (Deuteronomy 28:23), you will have a hard time finding work, advancing in your career, keeping a job, and so on. That is financial barrenness.

So by *hands* we speak of the hands themselves, as well as the use of those hands, **how** they are used. In the case of the heavy-handed, hurtful hairstylist, I'm sure I'm not the only one who stopped going there. See how that brings her business down, or to a halt. The same applies to the masseuse. Those are cursed hands.

Now, how do we know if a man's hands are under attack? Sudden loss of job and or income, for no apparent reason. Having the opposite of a green thumb for money. --anything you touch does not turn out well. Difficulty in gathering money. Anything you try to gather disappears. Difficulty in starting or growing, or even keeping a business. When one is very intelligent but has no job or any success, his hands may be cursed. The person who is working hard but has nothing to show for it may be suffering from cursed hands and therefore, barrenness.

But God says that Blessed shall be the works of your hand. When God curses a person, He curses the works of their hands.

Pray—

Let my hands be sufficient for me, (Deuteronomy 33:7)

My hands, do great things in the Earth, in the Name of Jesus.

Lord let every weakness in my hands be strengthened, in the Name of Jesus.

God empower my hands for good things.

God has destined that your hands will be empowered to accomplish great things.

Lord, You are the lifter of my head, and the lifter of my hands as with Moses in Exodus 17:11. Lord, send the helpers of my hands to strengthen my hands to assure me victory, in the Name of Jesus.

My hands, you are vital to my prosperity: my hands, be blessed and bless me, indeed, in the Name of Jesus.

Lord, teach my hands to fight and my fingers to war, in the Name of Jesus.

Lord, give me times of peace, so I can build. Let me build and not come down from this wall, in the Name of Jesus.

The enemy knows the importance of a man's hands and can wage war against it. Handcuffed, tied, padlocked hands – be set free, in the Name of Jesus.

Every arrow of failure fired into my hands, backfire, in the Name of Jesus.

Every attack on my hands, fail against me, and return to sender, in the Name of Jesus.

Resurrection power of Jesus come upon my hands now, in the Name of Jesus.

Lord, I repent, renounce, and denounce my own sins and every sin down my bloodline-

-, those I know about and those that I may not be aware of.

Initiations and evil marks can include incisions on the hands—including tattoos. You must pray against spiritual evil marks and remove physical evil marks whenever possible.

Curses such as barrenness can occur when rituals are done on the hands, or black, red, green, or white magic is done against a person using their fingernails for the ritual.

Lord, give me spiritually strong hands to receive blessings, hold blessings, share blessings-- hands that can transfer blessings, in Jesus' Name.

Lord, send deliverance to my hands, in the Name of Jesus.

Holes in my hands be sealed by the power in the Blood of Jesus. Hand paralysis, be healed, be made whole in the Name of Jesus.

Every curse against the work of my hands, backfire in the Name of Jesus.

Lord, I pray for every hand that will ever touch me, in the Name of Jesus: Therapeutic hands, dental hands, medical hands, healing hands, and helping hands--, let the Grace, Mercy and anointing of God flow through those hands and no other *spirit*, in the Name of Jesus.

Dry Hands

In the professional world, a doctor will have a lot of explaining to do if he or she has not practiced medicine or the discipline for which they have been trained for six months in most states of the United States. The term is to keep one's hands *wet*. This is a professional term that has been misused and misapplied online to other things. We are not talking about those other things.

Conversely, in the spiritual world, dry hands means that whatever the hands find to do, does **not** actually prosper. Procrastination could set in where work is never started. Or, projects could be abandoned, demolished, stopped, lost, or impossible to finish. Those hands are dry, so dry it is as though the hands are dead.

These next prayers are as prophesying to the hands to make those dry bones come together and live. Once alive, faith without works is dead, so those living hands should be doing *works* in the Earth. They should be doing what they were put here to do. When God blesses someone, He blesses the works of their hands.

Pray

My hands, get busy, get productive, in the Name of Jesus.

Most of the prayers, but not all, are adapted from prayers about hands by James Akanbi, from his **Blow Them by Thunder 3** *book.*

Lord, hear my repentance. Blood of Jesus reverse every curse on my hands, evident in the natural and in the spirit, in Jesus' Name.

Every generational idol, worship, or covenant causing *barrenness*, break today by the power in the Blood of Jesus.

Every satanic glove or covering preventing my hands from prosperity, flowing Fire like heavenly volcanic lava, turn them to ashes, in the Name of Jesus.

Every evil device inserted into my hands, be removed today by the Blood of Jesus.

Evil mark, incision, or tattoo on my hands be blotted out and my hand be made new again by the power in the Blood of Jesus.

All spiritual dryness on my hands, burn out today by Holy Ghost Fire, in Jesus' Name.

Every satanic hole in my pocket or in my hand leaking fortune and losing glory, be sealed by the power in the Blood of Jesus.

Spirit of death sent against my hands, see the Blood of Jesus; you must pass over me, you have no business here, no business with me, in the Name of Jesus.

Every satanic device or assignment that has been sent to kill my hands, I command you to fail against me, in Jesus' Name. You've been discovered; go to the pit: **Failed Assignment**.

My hands, be resurrected today by the Resurrection power in the Blood of Jesus.

Every covenant of dead hands and dead destiny, break, break, break, in Jesus' Name.

Every foundational power to delay my work or destiny, fail from today. You've been discovered; go to the pit for early torment, in the Name of Jesus.

Every strong man of my father's house to stop my progress by stopping the work of my hands, be paralyzed by Fire.

Every strongman of my ex, fake friend, enemy, relative, or evil stranger be bound and paralyzed by Fire, in the Name of Jesus.

Every power of touch attacking the work of my hands, be washed away today, completely, in the Name of Jesus.

Every evil hand upon my hand, upon my life, get off me, by Fire.

By the power of the God of Elijah, I cut off every evil hand against my destiny.

I declare that I am moving on, God is on my side and no man or woman can stop me in Jesus' Name.

Every demonic attacker planning to attack the work of my hand, *you* be attacked by the power of God in Jesus' Name.

Triple thread demonic spider assigned to stagnate through satanic cobweb, be neutralized, have no effect on me, and burn to ashes, in the Name of Jesus.

Every demonic poison in my hand be denatured by the anointing of God, in Jesus' Name.

Every spiritual drain connected to my hands, Thunder Hammer of God, crush it, in Jesus' Name.

Every arrow of the enemy in my hand, wasting my efforts, come out now; return 7-fold to sender, in Jesus' Name.

Every stubborn enemy trying to make me jobless, die, in the Name of Jesus.

Every evil effort by the curse of dead hands, cease by Fire, in the Name of Jesus.

Every financial frustration in my hands, stop now by the power in the Blood of Jesus.

All blessings, fortunes, and prospects falling off my business and life because of my hands, stop now, in Jesus' Name.

Every curse of stoppage or abandonment of my good works, be terminated today by the Blood of Jesus.

Every project in my hands that has stopped because of dead hands, be healed today by the resurrection power of Jesus Christ.

Every prosperous relationship that has been terminated by dead or cursed hands be renewed, revived, resurrected by the power of Jesus Christ.

Every hole anywhere in my life that is leaking prosperity and wealth, be sealed by the Blood of Jesus.

Anointing of God rest upon my hands now and propel my life forward, in Jesus' Name.

Holy Ghost Fire strengthen my hands today for success and prosperity, in Jesus' Name.

I command the West Wind of God to resurrect and restore my hands for profitable jobs, career and prosperity, abundance, and no lack, in Jesus' Name.

Hands, prosper by order of Zion, in the Name of Jesus. God takes pleasure in the prosperity of His servants. Amen.

My hands, be delivered today from barrenness, unfruitfulness, and non-productivity, in Jesus' Name. Amen.

Good Help Wanted

Yes, there can be evil curses levied against you, but a curse won't alight without cause. Yes, you could inherit a curse generationally through your bloodline, but the following are things you may be doing and not realizing that you are *creating* barrenness in your life and possibly increasing the iniquity of your bloodline, should you have a bloodline.

Working for the wrong person can bring on barrenness in your career, and life. Wasn't that the problem for Jacob working for Laban? Even though Laban was his uncle and ended up being his father-in-law, he lied to Jacob, cheated him and was a horrible boss. Laban also kept idols, so he was an idolater. Surely Jacob didn't expect to be dealt with in a Godly way by an ungodly man.

Working for a heathen can ruin your career and sometimes, your life. You need to learn as much as possible about a new employer before you accept a position in their company. Yes, this thing works both ways.

When I was an employee, on job interview days, I would say that I'm going to *interview the business,* because as far as I was concerned, God would give me the desires of my heart; if I wanted that job, as long as it wouldn't hurt me, or my destiny, God would give it to me. Now I had to do due diligence, and assess if the business was a proper company for me to work in.

Being in the wrong job, wrong occupation, wrong place, wrong time – being in rebellion, not doing what God has instructed you to do will not be fruitful for your life or your finances. Find out from God what career you should be in, sooner than later. Find out from God what city you should live in. Find out from God what company you should be working with or in. The scroll of your life was written by God. Find out from God if you are on your Destiny Clock right now, or if you are on a counterfeit clock supplied by the devil.

Employing the wrong person to work for you can bring barrenness to your business. In my case, the Lord deals with me quickly if I happen to hire the wrong person. It is sudden, it is drastic, and it is disheartening, but it is better to know sooner than later.

In my business the very first day the new employee starts, even though I may think I have vetted them well, and many times, if I am short staffed and short-sighted and only interview the applicant on paper and not spiritually, hoping for the best. This is not for the best; I have made errors. The Lord shows me my error their very first day: My business works by appointments, making them and by those appointments being kept. If there is a wrong soul in my camp, my telephone dries up for new appointments--, that day, immediately. Seriously, the phone just stops ringing. And I have had days where 20% to 50% of those on the schedule do not even come to their confirmed appointment just by hiring and having the wrong person on my staff. Hands cursed? Whole person cursed? A *desert, barren spirit*? Not sure, it is as though their presence repels productivity and success; that is not acceptable and will not work for me.

Good, bad, or indifferent as to their work skills, and experience, if they are **not** spiritually okay to work at my place, then they've gotta go. If God says they gotta go then, *bye Felicia*. So, I've implemented lengthy working interviews. By this process I can see the impact on my busyness without formally hiring the candidate.

You may feel sorry for this one or that one and hire them in your business that you have worked for some time to make successful. Our God is greater, but that may be a mistake. You must know their work history, work ethic, knowledge, and skill level. Are they qualified? Educated? Yes? Well, good. Also, can they pass a drug test, a credit check? Okay, good.

The credit check is almost the same in the natural as what you next need to ask God spiritually speaking, *Is this person under judgment from God? Is Heaven over this person shut up?* If they are under judgment from God, they may be little walking curses with cursed hands. Let's put it this way, would you hire Jonah if you knew Jonah was directly defying God? Of course not! You don't want to be collateral damage in Jonah's rebellion and mess. Recall how fast the fellows on the boat threw Jonah overboard once the tempest rose and was about to capsize the boat and kill all of them? Yeah, it's kinda like that.

Partnering with the wrong person, especially if God tells you not to, may be exactly what is bringing problems into your business and life. God hates evil alliances. Unless God sent you to a place, if you get in cahoots with someone that God is about to take down, you may somehow get on *their* wrong time clock and risk loss and or destruction. If the new person's destruction isn't imminent this may not be as noticeable as a storm rising.

As the business owner, every Friday or so, you will pay your employee or staff person. Whatever *altars* they worship at, your money is going there. In a real way, if they are workers of iniquity, if they are sinners, or dabble in the dark arts in any way, your money will go to altars that you know nothing about, or maybe you do know about them, but you think it means nothing because you don't believe in it, and so you think nothing will happen to you. Believe what you want, but when your business suffers or dries up, maybe you will believe differently.

There is a lot of spiritual warfare in business if you are going to do it God's way and be successful.

If you own an ice cream parlor and you hire someone who looks all Goth and scary, and then all their friends start coming there, or working there, don't you think they'd scare the kids away? And then the parents? And that's just on appearance, but most of the time unless the person is the cleverest of con artists, what they look like will be a sure reflection of **who they are on the inside**. If you are not a witch, why would you hire a staff of witches? Of course, no good person is going to dress up like a bad guy to try to trick another good person.

With conmen, a bad person will pose as a good person to get into your good graces, or into your business and end up ruining it accidentally or sabotaging it on purpose. Either way, you don't want the business you've put all your effort into to become a barren wasteland. *Right*?

Lazy Hands Will Be Barren

Lord, bless me indeed. Really? God blesses the hands of the diligent; when God blesses, He blesses the works of a man's hands. Barren hands could work, work, work, have 14 jobs, work 25 hours a day and have little to nothing to show for it. Don't let anyone judge or mischaracterize you if you are working diligently but being obstructed by barrenness.

On the other hand, if you are lazy, that will lead to nothing, disappointment and barrenness. Careful--, if you have a job and you are not doing it to the best of your ability, as unto the Lord, you are <u>creating</u> <u>*barrenness*</u> in your life. Faithful in a little, God will see you can be faithful over a lot. Being faithful over what is another man's, then God will give you your own. A willing and obedient man will eat the good and the fat of the land. Being a good worker sows that into your future, so when you are in charge and have people under you, *people shall be willing in thy day of power*, (Psalm 110:3a).

Every time you don't do what you're supposed to, what you agree to, what you are paid to do, instead, you are agreeing with unproductivity, you are agreeing with *laziness, procrastination, rebellion, barrenness,* and *poverty,* especially if you work for a Christian. When you agree with an idol *god*, or a demon long enough its an invitation and here they come into your life.

That Christian employer may not do anything to you regarding your employment. But a real Christian is praying and has a relationship with God and because of that, certain things are built in. if you work for a Christian but you're slacking off, hiding in the bathroom, playing on your cell phone, trifling about even coming to work, you are abusing one of God's people. Even if that Christian does nothing to you; eventually, God will have enough and God will handle it. It is a terrible thing to fall into the hands of an angry God. The last two people you'd ever want to cross is God and any one of *His*.

Even if you are not saved, the laws of being saved are in effect in a Christian business; don't work in one if you are just playing around, or because you think the boss will be nice to you. You will bring iniquity to yourself. Being unfruitful and unproductive will automatically bring a curse upon you. It will appear in your life in some aspect, as barrenness.

Perhaps that is how barrenness came into your life already; you inherited it from goofing-off, playing-around, get-over-minded ancestors. Pray God that you are not becoming one of those types yourself to pass barrenness on into your generations, if you get to have a family--, and ruin it.

In a whole other way, if you work for a person who is not saved, or worse maybe dabbling in dark arts, and you don't do what you agree to, you will incur ***their*** personal wrath in one way or another. So the lesson is, whatever you do, wherever you work, that's 8 hours in a day where you'd better do what you contracted to do, or there will either be built in iniquity, or hell to pay from the evil repercussions of a displeased, unsaved employer. You don't like anyone to mess with your business or money, *right*? The unsaved can take things to a whole other level. Fear not, but don't ***pray not***, either if you are the type to mess over someone's business or money.

How do I know this? The Holy Spirit told me.

God created the Heavens and the Earth and put man in the Garden to dress it. All that is raw materials. God gave man **hands--**, and with opposing thumbs to make use and make good of these raw materials.

Instead, a lazy man wants to lay back and have everything presented to him, as if God is *his* servant and the man is God. We need deliverance of the mind. Let this mind be in me Lord, that was in Christ Jesus.

Now, this is where the devil slides in. Either because he put that idea in man, or he can spot a lazy one who wants all the benefits of life to be handed to him, the devil promises this easy life. They are lies for the most part, but ready-made stuff is promised to a man in exchange for worshipping the devil.

If anything, ready-made comes into a person's life, by way of the devil, it was STOLEN, ready made from someone else that the devil had to kill or destroy to get it. Even though a Godly man may be making efforts to be productive and successful, using his hands, even blessed hands, he will appear and be as *barren* if he is being ripped off. Either way he does not enjoy the fruit of his labor, and money is a fruit.

Marital Barrenness

Where are the suitable suitors? And once you find them, how do you know if they are the right one, or not?

Too often barrenness in relationships or fruit of the body is the work of *spirit spouse*. I write extensively on *spirit spouse* in **Fantasy Spirit Spouse** and **Second Marriage, Third Marriage, any Marriage**.

Fruit of the body, fruit of the womb is addressed in Book 2 of this Barrenness series. It is entitled **Fruit of the Womb, Prayers Against Barrenness, Book 2.**

It Could Be My Fault

Barrenness in your life can result from sin, such as the shedding of innocent blood, taking bribes, stealing, even not returning the found property of others--, especially money. Repent.

Lord, forgive me for everything I have ever stolen, taken by mistake, found, *meant* to return, but did not--, borrowed, broke, ruined, kept, did not return, *meant* to return or repay, but did not.

Forgive me for credit card accounts that I have not paid, or any account not paid as promised that may be bringing *barrenness* to my life. Lord, I know You hate an unjust balance, broken covenants, and outright lies. Forgive me for every time I have not dealt honestly with people in business, or in my personal life in money, in any way, in the Name of Jesus.

Lord, have Mercy and remove all iniquity, in the Name of Jesus.

Lord, I can see now in all the ways I brought iniquity on myself, even with good intentions. The road to hell is paved with good intentions, Lord, have Mercy.

Help me to follow through and complete things I start. Help me start things I should start. Stop me from starting things I should not start and help me finish things I should finish. Lord, help me reach success, victory, and destiny, in Jesus' Name.

Lord, help me to be organized, and keep up with things. Anything borrowed that needs to be returned, help me return it, in the Name of Jesus.

Wealth Transference

Pray

The wealth of the wicked is laid up for the just. Lord, let me be among your covenanted justified that you will transfer to me in the Earth from those whose job it is to heap up and pile up, in Jesus' Name.

Lord, also, as indicated, transfer to me from my heavenly stores and Your own treasures that which is needed for life and for godliness while here in the natural, in the Name of Jesus.

Power of the Blood of Jesus separate me from the sins of my ancestors. I renounce any evil dedication placed upon my life, in the Name of Jesus.

Lord, let evil vows against my future be rendered null and void, in Jesus' Name.

You, powers that quench the Fire of God in my life, I am not your candidate, in the Name of Jesus.

Breakthrough Angels of God, assist my Angels of Blessings against every satanic hindrance, in the Name of Jesus.

Every sickness and disease, come out with all your roots now; Lord, give me strength to persevere, strength to work, and strength to run on to see what the end will be, in the Name of Jesus.

Lord, deliver me from evil stones thrown at me by unfriendly friends, exes, jealous competitors, and evil human agents--, household, or strangers, in Jesus' Name.

My head will not be anchored to any evil, in the Name of Jesus.

Let evil pursue all unrepentant evil workers that are pursuing me, in the Name of Jesus.

Every delay against my progress fall down and die, in the Name of Jesus.

I reject satanic restriction in every area of my life, in the Name of Jesus.

God of Abraham, Isaac, and Jacob, manifest Yourself as the Covenant-Making God, in the Name of Jesus.

Every power challenging the power of God in my life, die, in the Name of Jesus.

Let every satanic plan against my future cease and desist, in the Name of Jesus.

Spirit of failure, be disgraced, in the Name of Jesus.

Spirit of impossibility leave me; all things are possible with God. I declare against you: failed assignment, in the Name of Jesus.

Spirit of giving up or *giving in*, I'm not your candidate; you give up and go to the pit for early torment, failed assignment against me, in the Name of Jesus.

Spirit of barrenness, spirit of unproductivity, not accomplishing or finishing anything, be bound and removed from my life right now, in Jesus' Name.

Every *desert spirit* coming against me, be destroyed in the Name of Jesus.

Spirit of debt, poverty, insufficiency, lack, and bankruptcy, against me, be paralyzed now, in the Name of Jesus.

I reject every *spirit of business*, work and *career failure,* in the Name of Jesus.

Spirit of destruction of good things in my life, be paralyzed now, in the Name of Jesus.

Desert spirit, dry up completely by the Fire of the Holy Ghost, in the Name of Jesus.

Blood of Jesus, block poverty in my life, in the Name of Jesus.

Every desert strongman assigned against my life, fall down and die now, in the Name of Jesus.

I paralyze the activities of *desert spirit* in my life, hit the dirt, and die, in the Name of Jesus.

Every evil load of *desert spirit* in my life, go back to your sender, 7-fold, in the Name of Jesus.

All powers assisting, associated with, or accompanying poverty in my life, be bound, be paralyzed, be ineffective, be cast out of my life, in the Name of Jesus.

My life, receive the anointing of fruitfulness, and productivity, in the Name of Jesus.

Desert spirit: every *spirit* that keeps a person in barrenness, I send the East Wind of God to destroy you to the utmost, DIE! DIE! DIE, in Jesus' Name.

Cursed Legs

When you are forever in the wrong place at the wrong time or you just seem to stumble upon trouble and problems, you could have *cursed legs*. This is another avenue to *barrenness*. To feel something touching or pulling your legs in the dream or in the night means that a curse has been placed upon your legs. This type of curse will tend to poverty. Poverty and barrenness work together as you can see. As in cursed hands, to have cursed legs means you need to pray as well for deliverance of the head as well as the legs.

If you have sudden leg, ankle, foot pain, you went to bed fine and just woke up like that one morning –suspect witchcraft attack of the legs. If you have leg troubles, or get bruised or injured legs often, suspect witchcraft against you. Of course, do your due diligence. Be sure you are healthy and exercising, sleeping in proper positions. Lower extremity pains can be a result of compressed nerves. Do not suspect a witchcraft attack if you're sleeping on the couch in fetal position night after night. That pain is a result of your own doing.

Lord, deliver my head, in the Name of Jesus. Every head malady, cease, be healed, in the Name of Jesus.

Every hand of affliction be taken off me now, in Jesus' Name.

Every affliction of the legs and ankles be removed far from me, in Jesus' Name. *Amen.*

The following prayers against cobweb attack of the legs, feet, ankles prayers are excerpted from my book, **Prayers Against Demonic Cobwebs.**

I nullify and neutralize the magnetic power of every demonic cobweb, whether I come into contact with it or not, I render it neutral, useless, powerless, and harmless against me, in the Name of Jesus.

Lord, let the intent of every cobweb that has been sent against me in my household, my business, my life, and the life of my family, backfire! Return to sender, in the Name of Jesus, by the power in the Name of Jesus.

I decree and declare total destruction of all evil powers using demonic cobwebs that have vowed to keep me in the same job, same house, same place, trudging along the same path and career for the rest of my life without progress, in Jesus' Name. You're a liar; I shall progress and reach the destiny that God has planned for me from the Beginning.

By the power in the Name of Jesus, let every demonic cobweb break into irreparable pieces, in the Name of Jesus.

Father, send Your Warrior Angels with swords drawn, to locate and completely destroy all demonic cobwebs on any doors, rooms, roads, windows, public places, and any other spaces where these wiry traps have been set to trap me or my family, in the Name of Jesus.

By the power in the Name of Jesus, I decree and declare destruction of every demonic cobweb of poverty, sickness, infirmity, backwardness, setbacks, stagnation, disappointment, confusion, hatred, rejection, reproach, non-achievement, barrenness, failure, and any covering over my life, myself, or my family, in the Name of Jesus.

Let every demonic power that is using spiritual cobwebs to steal from me, store, and eventually divert my God-given blessings to a demonic storehouse, fall down and die, in the Name of Jesus.

By the power in the Name of Jesus, I decree and declare that whosoever would try to reactivate these demonic cobwebs and spider stings that have hindered my ancestors and my parents from succeeding in life, let them be destroyed by Fire, in Jesus' Name. Return to sender!

My legs, break free of every cobweb trap, do your job, take me forward toward destiny, in the Name of Jesus.

My legs, take me to my destiny helper, in the Name of Jesus.

My legs, take me to the place of my miracles and where God has commanded the blessings. Amen.

Let this mind be in me that was also in Christ Jesus; head, hands, feet, legs, obey the Word of the Lord, in Jesus' Name.

Bad luck, disfavor: **cease** against me, in the Name of Jesus.

Anointing fall on me and give me favor in the eyes of men, angels, and God, in the Name of Jesus.

Every spell, curse, mischief, hex, vex, incantation against me: return to sender, in the Name of Jesus.

Lord, You are the lifter of my head, in the Name of Jesus.

Every evil touch on my leg – every power involved, keep your hands to yourself and die, die, die, in the Name of Jesus.

You will not sow *poverty*, sorrow, or *barrenness* into my life, in the Name of Jesus; return to sender.

Fire Back

Pray

I fire back every witchcraft arrow fired at my head, in the Name of Jesus.

Powers of my father's house against my future, destiny, and glory, fall back and die, in the Name of Jesus.

Serpents and scorpions assigned against my head, die, in the Name of Jesus.

I reject the *spirit of the tail* and I claim the *spirit of the head*, in the Name of Jesus.

Let every incantation, incision, hex, or vex working against me backfire to sender, in the Name of Jesus.

I break all satanic connections and any linkage to strange people, in the Name of Jesus.

Lord, turn away all that will jilt, disappoint or fail me, in the Name of Jesus.

Finger of God, pull down every stronghold up against me, in the Name of Jesus.

Holy Spirit redirect every evil wind sent against my life, in the Name of Jesus.

Lord, forgive me and I take authority over curses emanating from evil dedication.

Ancestral Wickedness

Ancestral sins have iniquity, and that iniquity is automatically inherited if the sin is not repented of, and the iniquity removed from the bloodline.

What anyone's ancestor did, only the Holy Spirit knows and will tell you truthfully where you don't need to doubt if half of it is a lie or not. Your ancestors could have handled cursed objects. They could have practiced in the occult as an evil human agent of Satan, or any other of a number of things. They could have done this knowingly, or they could have been tricked, being spiritually ignorant. Iniquity still follows.

Living in a cursed land or housing, will bring on troubles and limit the flow of God's blessings to you. Being born into a satanic family, using strange money, or accepting cursed gifts, partaking of food sacrificed to idols--, consciously or unconsciously still honors demon idols. Conscious or unconscious performance of demonic rituals are nothing to be laughed at. Every festival you go to, if it is not honoring God, it is honoring and giving worship to an idol *god*. Just going to some of these places constitutes initiation of worship and covenant with idol *gods*, and you thought you were just having fun and some funnel cake. Once there, know that the level you *participate* at the festival is the level of your involvement with the idol *god* whose festival you attended. In this way, people **do what they know not.** (Luke 23:24)

Pretty much every secular music concert you've been to has a pack of *spirits* in there and many times *spirits* are channeled through those mics and speakers. What idol or demons is the artist channeling? It could be many. It depends on what that person's deal is with the devil, if they indeed have one. They might have one and not even know it.

I break every curse which may be in my parent's families back to Adam and Eve, on both sides, in the Name of Jesus.

I take authority over every curse in my family line--, every curse of sickness, disease, deformity, destruction, financial upheaval and disasters, broken marriages, fear, physical and spiritual destruction. I cancel the consequences and evil effects of all curses against me, in the Name of Jesus.

I cancel every evil initiation that I have entered into knowingly or unknowingly, in my awake life, or in the dream, in the Name of Jesus. Lord, forgive my ignorance for going to festivals, parades, parties, fairs, concerts, and other revelries with abandon and without knowledge, in Jesus' Name.

I take authority over every unconscious, blind curse issued against me; Lord, purge the root of my life with Holy Ghost Fire. Lord, don't let me pass these curses on to my children, in the Name of Jesus.

I break every financial, health, or beauty curse placed on me out of jealousy, or any other reason, in Jesus' Name.

I break and cancel every clinical and medical curse, in Jesus' Name.

I break and cancel every curse issued by satanic ministers, in the Name of Jesus.

I break and cancel every curse of exchanged destiny, in the Name of Jesus.

I break and cancel every curse emanating from evil prophecies by false or fallen prophets and *lying spirits*, in Jesus' Name.

Lord, have Mercy, forgive me and release me from evil and negative words out of my own mouth against You or Your Spirit. Forgive me for tithe failures. Lord, I am sorry, forgive me for breaking anyone's heart. Forgive me for disobedience, rebellion, deliberate sin, and self-imposed curses.

Lord, I renounce and denounce involvement in false religions and cults.

Lord, forgive me where I have overstepped Your Word and instruction in any way.

Forgive me for sexual immorality, perversion, for conscious and unconscious demonic sacrifice.

I reverse all curses, Lord. There shall be no more sickness, poverty, delay, illness etc. and barrenness in my life, in the Name of Jesus. I command it.

Bless the Lord, oh my soul, and all that is within me. Bless His Holy Name.

Loose All Bondage

In my book, **Level the Playing Field,** healing of the foundation is dealt with at length. It is very important for this topic as well, for that reason, I do recommend you read it.

Lord, thank You that you have the power to deliver me from any bondage, in the Name of Jesus.

I confess my sins and sins of ancestors, especially those linked to evil powers.

Lord, send Your axe of Fire to the foundation of my life and destroy every evil plantation.

I vomit every evil consumption, all spirit food that I have been fed from a baby to now. I vomit it with projectile force and heavenly acid to burn and completely destroy those who fed it to me in the dream, in the night, in the Name of Jesus.

I command all foundational strongmen attached to my life to be paralyzed, in the Name of Jesus.

Lord, strike down any rod of the wicked against me and my family, in the Name of Jesus.

I reject and renounce any evil name given to me consciously or unconsciously in any evil association, in the Name of Jesus. The Lord knows my name and that is the only name I will answer to, in Jesus' Name.

I cancel the consequences of any evil name attached to me; any name that will not allow my star to shine, die, in Jesus' Name.

Problems transferred to me from my mother's womb, I cancel you, by the power in the Blood of Jesus.

Lord, cleanse every organ of my body by the Blood of Jesus and the Fire of the Holy Ghost, in Jesus' Name.

I break every inherited evil covenant in Jesus' Name.

I break every inherited evil curse, in the Name of Jesus.

I bind and paralyze every demon sent to enforce the curse, and I cast them into the pit, in the Name of Jesus.

Lord, destroy the iniquity of in my foundation, consequences of it in my past present and future relationships, in the Name of Jesus.

Lord, I pray against every parental curse, every evil and household rivalry, evil dedication, fellowship with idols, demonic incisions, spiritual marriages, dream defilement and pollution, bewitchments, the evil laying on of hands, demonic sacrifice, idols of my father's house, demonic initiations, inherited diseases, disorders or syndromes, wrong exposure to sex, exposure to evil diviner(s), demonic blood transfusion, in the Name of Jesus.

Lord, transfuse my Blood with the Blood of Jesus.

Let every gate opened by the enemy to my foundation be closed and sealed forever from demonic access, by the Blood of Jesus.

Lord, Jesus, deliver me, heal me, and make me over to be as You created me in Your image and likeness, in the Name of Jesus.

I reject, revoke and renounce any membership with any evil association, cult, or secret society, in the Name of Jesus.

Lord, deliver me from all the word curses that I have spoken out of my own mouth. Deliver my mouth of unclean lips, in the Name of Jesus.

Forgive me for all the times I said words of doubt, unbelief, anti-success--, any negative words, any words of failure, over my own body, life, marriage, children, job, or business, in Jesus' Name.

I renounce and revoke all the oaths I took consciously or unconsciously for any reason, at any time, in any place, in any dimension, realm, timeline, or age, including oaths in the dream that have been wiped from my memory, in Jesus' Name.

I break and cancel every evil mark, incision, writing placed in my spirit and body as a result of my membership of any evil, occultic association, in Jesus' Name.

I withdraw any of my body parts, blood or other bodily fluid in the custody of any evil altar, in Jesus' Name.

I withdraw my pictures, image, any part of my body, my soul, and my spirit from every evil altar, in the Name of Jesus.

I return any of the things of evil associations I am consciously or unconsciously connected with. I declare, I do not want it, and I am not in covenant with any evil entity or evil power, in the Name of Jesus.

I withdraw and cancel my name from every evil register by the Blood of Jesus, in the Name of Jesus.

Holy Spirit, build a wall of Fire around me that will make it impossible for any evil *spirit* to come at me again.

Lord, break down every evil foundation of my life and rebuild a new one on Christ the Solid Rock.

I break every curse placed on me by my parents, purposefully, or in ignorance, in the Name of Jesus.

I cancel and break every curse, spell, hex, enchantment, bewitchment, incantation placed upon me by any satanic agent, at any time of my existence, in the Name of Jesus.

Fire of God, roast and burn to ashes every evil bird, snake, spider, gecko, or other animal sent into my life, in Jesus' Name.

I break, cancel, and dismantle every inherited curse, in the Name of Jesus.

I dismantle every hindrance, obstacle or blockage put in the way of my progress by any evil association, in the Name of Jesus.

I break and revoke every blood and soul tie covenant and yokes attached to any satanic agent, in the Name of Jesus.

Doors of blessings and breakthrough shut against me by any evil association, I command you to be opened, in Jesus' Name.

Lord, thank You for permanent deliverance, let these same pursuers never catch or pursue me again, in Jesus' Name.

I break myself loose from every collective covenant, in the Name of Jesus.

Let all evil competitors tumble and fall, in the Name of Jesus.

Lord, let all my adversaries make mistakes that will cause them to fall and be captured, and advance my cause, in Jesus' Name.

I break and cancel every evil covenant with idols and all attached yokes, in the Name of Jesus.

I send confusion into the camp of all evil agents planning against my fruitfulness, in the Name of Jesus.

I break and cancel every evil covenant my parents entered into, whether they thought they were doing right, or not. I break every yoke attached to it, in the Name of Jesus.

I command darkness and confusion upon the camp of the enemy, in Jesus' Name.

I remove my name from the books of barrenness, unfruitfulness, non-productivity, poverty, and failure and demonic sidetracking, in the Name of Jesus.

I declare my body, soul and spirit are a no-fly zone for all evil *spirits,* in the Name of Jesus.

Fire of God, roast the forces of hindrance and obstacles and paralyze their powers, in the Name of Jesus.

Lord, give me power to make use of divine opportunities. Let me possess more knowledge and Wisdom than my competitor, in Jesus' Name.

Make my paths confusing to the enemy, Lord God.

Let me always be ahead of my competitors in business and in life, in Jesus' Name.

Let all the adversaries of my breakthrough be put to shame, in the Name of Jesus.

All *scorpion spirits* sent against me, lose your sting, lose your sting, lose your sting against me, in the Name of Jesus.

Every power circulating my name for evil, receive Holy Ghost slap and disgrace, in the Name of Jesus.

Blessings of Abraham

Pray

I sing praises and worship to You, O Lord. Praise the Lord for His mercy endureth forever.

Blessed shall be the fruit of the ground-- Thank You Lord, for making us plenteous in goods, and our careers, in the land which You gave us, according to the covenant made with our Fathers, (Deut 28:4, 11)

The blessings of Abraham come upon us, through Jesus Christ, in the Name of Jesus, (Gal 3:13-14)

I break every covenant with any demon, or idol *god* making counterfeit promises to me; my blessings come from God and from God, alone, in Jesus' Name.

I breathe in the Fire of God, and I breathe out all negative things, in Jesus' Name.

Thank You Lord, for provision; You are the Lord.

Let all satanic banks that are harboring my blessings be paralyzed, in Jesus' Name.

Lord, I pray against unfriendly friends, evil relatives, and stealth enemies.

I pray against demonic initiation through any means, be it the evil laying on of hands, food in the natural or spirit food, by the Blood of Jesus.

Household Witchcraft

Household witchcraft is among the most dangerous. People closest to you who know you and know your secrets and weaknesses and are secretly working against you is household witchcraft. No matter what level they take it to, if it is against you and against what God has for your life, it is witchcraft.

Thunder of God locate and dismantle the throne of witchcraft in my household, in the Name of Jesus.

Every seat of witchcraft in my house and household be roasted with the Fire of God, in the Name of Jesus.

Thunder of God, scatter beyond redemption the foundation of witchcraft in my household, in my bloodline, in the Name of Jesus.

Every stronghold or refuge of household witches, be destroyed, utterly, in Jesus' Name.

I break free from every bondage of witchcraft covenant, in the Name of Jesus.

Any place where my blessings are hidden, catch fire, and roast to ashes, by the Fire of God, in the Name of Jesus.

I frustrate every plot, device, scheme, and project of witchcraft designed to affect any area of my life, in the Name of Jesus.

Any organ of my body that has been removed or exchanged through witchcraft be restored, now, in Jesus' Name.

Household wickedness against my success in life, stand down. (X3) or receive the wrath of God.

Deliverance & Destiny Helpers

Lord, hear me in my trouble and let the name of Jacob defend me. Send me help from the sanctuary and strengthen me out of Zion, in the Name of Jesus. (Psalm 20:1)

I jump out of my generational bloodline and curses into the bloodline of Jesus Christ, by adoption, in the Name of Jesus.

I renounce any evil dedication placed upon my life, in Jesus' Name.

I renounce and break every evil ordination, and dedication placed on my life, in the Name of Jesus.

I command all demons associated with evil dedication to leave now, in Jesus' Name.

Lord, cancel the evil consequences of any broken demonic promise or dedication.

I take authority over all the curses emanating from breaking the vows made during any evil dedication, in Jesus' Name.

I command all demons associated with any broken evil parental vow and dedication to depart from me now, in the Name of Jesus.

Lord, separate me completely from all the sins of my forefathers by the precious Blood of Jesus. Lord, remove the curse if it is from You.

I cancel all ungodly delays to the manifestations of my blessings, and miracles, in the Name of Jesus.

Every evil touch on my head or hand, I break your power; intended curse, go back to sender, 100-fold, in Jesus' Name.

I reverse all curses, Lord. There shall be no more sickness, poverty, delay, illness etc. and barrenness in my life, in the Name of Jesus. *Amen.*

Angelic Assistance

Lord, send angelic assistance, in Jesus' Name.

Let the angels of the Living God roll away every stone of hindrance to my breakthroughs, in the Name of Jesus.

O Lord, hasten Your Word to perform it, in the Name of Jesus.

O Lord, avenge me of my adversaries, in Jesus' Name.

I refuse to agree with the enemies of my progress, in Jesus' Name.

My hands, come out of captivity, now. Break free of every fetter, chain, iron, rope, padlock, anything tying you together, in Jesus' Name.

My hands, be fruitful, be prosperous, multiply like Jacob and Joseph, in the Name of Jesus.

Lord, let **Your** plans for my hands go into full effect, in the Name of Jesus.

Kingdom work, prosper in my hands, in the Name of Jesus.

Lord, hear my cry, and deliver me; deliver me speedily; You are my strong rock and my defense; You are the Lord.

Angels of God interfere with any power, force, entity, or agent attempting to interfere with the good works of my hands, in the Name of Jesus.

Father, send out your angels to unearth and break all evil storage vessels fashioned against me, in Jesus' Name.

In the Name of Jesus, send Your mighty angels to *loose* me from all demonic holds, psychic powers, bonds of physical illness, and bondages, in the Name of Jesus.

My angel of blessings, locate me today, in the Name of Jesus.

Mighty hands of God be upon me for good, in Jesus' Name.

God, arise, and be God against my oppressors, in the Name of Jesus.

Prosperity in finances, health and marriage, in everything I do, come into my life, home and relationship today, in Jesus' Name.

Lord, restore me to the way You created me, if I've been altered, in Jesus' Name.

Lord, give me my Rehoboth, the land where God has made room for me, for marriage, for victory, success, and destiny, in Jesus' Name.

The Thieves

Thieves come to plunder, to steal, kill and destroy. They come to strip fruitful lands leaving them naked and barren.

The Lord commands His East Wind which can bring everything that can ravage a land and leave it desolate. Sadly, evil people have learned this and if we leave the enemies of God to do as they please, they will use divine weapons *against* the people of God. Do not let that happen! Do not let evil stand or be the last word that the winds or any of the other elements hear, daily!

Sin is a thief. Personal sin can cause judgment to come upon a person, a people, even a land. Even if all of you got together and voted on something that God didn't approve of, but you all decided it was good, judgment will come anyway if it is not God approved--, upon people and sometimes upon the land, as well. God is not mocked.

I take back everything the enemy has stolen from my life; Devil, you will not steal blessings from my life again, in Jesus' Name.

I fire all satanic bankers and managers of satanic banks, operating against me, in Jesus' Name.

Thunder of God, break into pieces satanic strongrooms holding my stuff; I bind and plunder every strongman, in Jesus' Name.

Everything the enemy has taken from me, I take it back by force and by Fire, in the Name of Jesus.

I possess my possessions, in Jesus' Name.

The God of Elijah who answers by Fire, answer me by Fire, in the Name of Jesus.

My hands, I command you to prosper in all areas of my life, in Jesus' Name.

My hands, receive the strength of God and prosper, in the Name of Jesus.

Angel of the Living God, revive my hands today, in the Name of Jesus.

Lord, destroy by Fire whatever interferes with Your promises to me, in Jesus' Name.

Let the Blood, and the Fire destroy very wicked work against my productivity in the Earth and in the Spirit, in Jesus' Name.

Let every area of my life become too hot for any evil, in the Name of Jesus.

I reject all evil manipulations and manipulators, in Jesus' Name.

Witchcraft and *familiar spirits* over my life, break, in Jesus' Name.

Lord, permanently dislodge and dismantle the power of the enemies that are up against me, in the Name of Jesus.

Earth, O Earth, open up and swallow up all my enemies, in the Name of Jesus.

Every bewitchment, and curse against me, lose your power, authority, and legal standing to work against me; I plead the Blood of Jesus as my defense.

I am the righteousness of God in Christ Jesus; you have no power over me.

Arrow of *barrenness* fired against me, back to sender, in Jesus' Name.

Good success, appear in my life, permanently, in Jesus' Name.

Lord, thank You for Wisdom to use the knowledge You've given, in Jesus' Name.

Seed of barrenness satanically sown in my soul to provoke barrenness in my life, be removed by the Blood of Jesus.

Strangers from the waters troubling my dream life, fall flat on your face, and die, in the Name of Jesus.

Power In the Hands

Holy Ghost power, fall on me now.

Father, bless my hands, give strength and anointing for great achievements, all to Your Glory, Amen.

Lord God release Your blessing on my hands, so whatever I set my hands to, prospers, in the Name of Jesus.

Power of God, empower my hands to do good works and exploits in the Earth, in the Name of Jesus.

I receive the power to get wealth by the Word of God.

I take up the power to receive wealth by the power of God.

I receive the power to hold wealth and not be robbed; the devourer is rebuked from my life, in Jesus' Name.

I receive the power to enjoy the fruit of my labor, in Jesus' Name. Money is a fruit.

I receive the power to share and give wealth where appropriate, by the Word of God, in the Name of Jesus.

I receive Wisdom to not give wealth at wrong times, in wrong places, or to make unholy alliances, in the Name of Jesus.

I receive the power to transfer wealth to my generations, appropriately, at the right time, in the Name of Jesus.

I bind every obstinate, vengeful, retaliatory *spirit* and command all backlash against me because of these prayers to backfire, in the Name of Jesus.

I seal these declarations across every age, realm, dimension, and timeline, past, present, and future, in the Mighty Name of Jesus Christ. **AMEN.** *Amen.*

Christian books by this author

AK: Adventures of the Agape Kid

AMONG SOME THIEVES

Ancestral Powers

As My Soul Prospers

Behave

Churchzilla (Wanna-Be Bride of Christ)

The Coco-So-So Correct Show

Demonic Cobwebs

Demonic Time Bombs

Demons Hate Questions

Do Not Orphan Your Seed

Do Not Work for Money

Don't Refuse Me Lord

Every Evil Bird

Evil Touch

The FAT Demons

Fruit of the Womb, Prayers Against Barrenness,2

got Money?

Let Me Have a Dollar's Worth

Living for the NOW of God

Lord, Help My Debt

Lose My Location

Made Perfect In Love

The Man Safari *(I'm Just Looking)*

Marriage Ed., *Rules of Engagement & Marriage*

Motherboard: *Key to Soul Prosperity*

Name Your Seed

Plantation Souls

The Poor Attitudes of Money

Power Money: Nine Times the Tithe

The Power of Wealth (forthcoming)

Prayers Against Barrenness, Book 1

Seasons of Grief

Seasons of War

Second Marriage, Third Marriage any Marriage

SOULS in Captivity

Soul Prosperity: Your Health & Your Wealth

The *spirit* of Poverty

This Is *NOT* That

The Throne of Grace, *Courtroom Prayers*

Warfare Prayer Against Poverty

When the Devourer is Rebuked
The Wilderness Romance

Other Journals & Devotionals by this author:
The Cool of the Day – Journal
got HEALING? Verses for Life
got HOPE? Verses for Life
got WISDOM? Verses for Life
got GRACE? Verses for Life
got JOY? Verses for Life
got LOVE? Verses for Life
He Hears Us, Prayer Journal
I Have A Star, Dream Journal
I Have A Star, Guided Prayer Journal,
J'ai une Etoile, Journal des Reves
Let Her Dream, Dream Journal *in colors*
Men Shall Dream, Dream Journal,
My Favorite Prayers (in 4 styles)
My Sowing Journal
Tengo una Estrella, Diario de Sueños
Illustrated children's books by Dr. Miles

Big Dog (8-book series)

Do Not Say That to Me

Every Apple

Fluff the Clouds

I Love You All Over the World

Imma Dance

The Jump Rope

Kiss the Sun

The Masked Man

Not During a Pandemic

Push the Wind

Tangled Taffy

What If?

Wiggle, Wiggle; Giggle, Giggle

Worry About Yourself

You Did Not Say Goodbye to Me

www.ingramcontent.com/pod-product-compliance
Lightning Source LLC
Chambersburg PA
CBHW061336040426
42444CB00011B/2945